Sč̓iłpálqʷ

Biocultural Restoration of Whitebark Pine on the Flathead Reservation, Montana

Sč̓iłpálq^w

Biocultural Restoration of Whitebark Pine on the Flathead Reservation, Montana

Michael Durglo, Jr., Richard G. Everett, Tony Incashola, Jr., Maureen I. McCarthy, ShiNaasha H. Pete, Joshua M. Rosenau, Séliš-Ql̓ispé Elders Cultural Advisory Council, Thompson Smith, Shirley Trahan, and Anne A. Carlson

Published by
Salish Kootenai College Press
Pablo, Montana

Distributed by
University of Nebraska Press
Lincoln, Nebraska

Cover illustration: CSKT Climate Change Advisory Committee Chairman, Mike Durglo Jr., with ił̓awy̓eʔ (Eel-lah-wee-yeh) or "great-great-grandparent." Photo credit: Mike Durglo Jr.

Library of Congress Cataloging-in-Publication Data:
Names: Durglo, Michael, Jr., author
Title: Scilpalq: biocultural restoration of Whitebark pine on the Flathead Indian Reservation, Montana / Michael Durglo, Jr. (and 7 others).
Other titles: Scilpalqw, Whitebark pine
Description: Pablo, Montana : Salish Kootenai College Press, (2025) | Originally published in: Climate actions, edited by Brenda Groskinsky, published in 2022 by CRC Press. | Includes bibliographical references and index.
Identifiers: LCCN 2025017723 | ISBN 9781934594360
Subjects: LCSH: Whitebark pine--Regeneration--Montana--Flathead Indian Reservation | Forest restoration--Montana--Flathead Indian Reservation
Classification: LCC QH105.M9 D874 2022 DDC 333.75/1530978682--dc23/eng/20250721
LC record available at https:/Lccn.loc.gov2025017723

published by
Salish Kootenai College Press
PO Box 70
Pablo, MT 59855

Distributed by
University of Nebraska Press,
1225 L Street, Suite 200
Lincoln, NE 68588-0630,
order 1-800-848-6224, www.nebraskapress.unl.edu.

Table of Contents

Sč̓iłpálqʷ: Whitebark Pine and the Séliš and Ql̓ispé People[1]

From the beginning of human time, the people of the Séliš (Salish or "Flathead") and Ql̓ispé (upper Kalispel or "Pend d'Oreille") nations have lived in the valleys and mountains of the Northern Rockies and adjoining areas (SQCC 2021). Across these vast Indigenous homelands, the ancient trails follow the ridgelines and crisscross the mountain passes. Tribal people have gone to those places for many reasons, including gathering foods and medicines, hunting and making prayers and offerings. And the people have always known the tree that grows only there. In the Salish language, it is called sč̓iłpálqʷ. English speakers refer to it as whitebark pine. In scientific terminology, it is named *Pinus albicaulis*.

Whitebark pines can live for more than a thousand years. That's twice the five centuries since Columbus first landed in the Americas and initiated the transformation of the continent. But it is still a small fraction of the 12,000 years, or more, that the Séliš and Ql̓ispé have lived with sč̓iłpálqʷ. The tribes' collective memory holds a deep knowledge of many generations of whitebark pine, and how they have stood present, watching over the people from the ridgetops.

In this book, we describe the efforts of the Confederated Salish and Kootenai Tribes (CSKT) to save and restore sč̓iłpálqʷ, which is now threatened with extinction from accelerating and intensifying effects of the climate crisis and disease. Too often, there is too little communication and exchange between Indigenous communities working on issues of cultural survival and restoration and scientists focused on their own research methodologies and approaches. On the Flathead Reservation, however, the disparate worlds are being brought together in a visionary effort to save whitebark pine, reflecting what is known as biocultural restoration.

Biocultural restoration on the Flathead Reservation goes beyond simply combining "culture and science" in a simplistic way. Building on this community's long history of educational projects that have bridged these disciplinary divides (CSKT–NRD 2021b, 2021a, 2021c), we are searching for, and finding, ways to ensure that it is the tribes' cultural relationship with sč̓iłp̓álqʷ that structures what we do and how we apply the powerful tools of Western science. This is not another case where the superficial trappings of Indigeneity are used as little more than a colorful logo on an otherwise conventional scientific methodology. Instead, tribal cultural ways are shaping every aspect of this endeavor, from the purpose and goals of the project to how we physically engage with the trees. We are taking care to ensure that scientific perspectives do not redefine our relationship with this tree, but rather that we are employing science as a tool to pursue cultural objectives.

Before we describe our scientific work, let us offer a brief encapsulation of the Séliš-Ql̓ispé cultural relationship with sč̓iłp̓álqʷ. The Séliš and Ql̓ispé have known sč̓iłpálqʷ in ways that are hard to fully explain to the public. It has to do with both the time-depth of the relationship, and also how the tribes have lived and prayed, how every aspect of the way of life has been centered in a deep connection with the plants and animals. In Séliš-Ql̓ispé creation stories, the plants and animals make themselves into food and medicine. In giving of themselves, they make possible the people coming into this world, providing sustenance and life. The elders often remind us that we are forever indebted to plants and animals for their sacrifice. At the heart of Séliš-Ql̓ispé cultures, reflected and reinforced in the prayers and ceremonies, human beings occupy a place of humility, relying upon the material and spiritual help of plants and animals. In return, people are obligated to always treat plants and animals with respect. These relationships are full of joy and happiness in the gifts the people receive each day.

The elders have said that it is from those creation stories, from the sense of gratitude to plants and animals, that the people must honor the plant or animal, the one who gave its life so that the people might live. So, the ethic of avoiding waste stood at the center of the traditional way of life. As Ql̓ispé elder Pete Beaverhead said in a 1975 recording, that the ethic held true

> "…for anything they gathered or killed", "Kʷem̓t pentč u esyaʔ u es čteʔm̓m łu sp̓iqałq u łu sw̓ew̓ł̓ u łu st̓at̓ap łu skʷiskʷs łu stem̓. Esyaʔ u es ʔiłistm̓. Esyaʔ u es čteʔm̓m. (They always used all of the berries, the fish, the things they killed, like the ruffed grouse or anything else they gathered or killed. They ate all of it. They used all of it.) [P. Beaverhead, in Séliš-Ql̓ispé Culture Committee archives (SQCC 2021)]

In another 1975 recording, Séliš elder Louise Wisqn Vanderburg conveyed the same central principle, "Šey̓ ɫu sʔiʔiɫis łu tsq̓si … Esyaʔ še kʷeʔeys. Ta epł es x̣ʷel. ("That was their food a long time ago…They took everything. They didn't throw anything away.") [L. W. Vanderberg, in Séliš-Ql̓ispé Culture Committee archives (SQCC 2021)].

The elders have told stories about sc̓iłpálqʷ. They remembered camps up in the mountains, near places like Čpáaqn, the mountain on the Reservation Divide. They told how the people would gather some of the heavy, dense cones and place them by the fire, where they roasted them until they opened up. The cones were full of seeds, called sc̓eyłp. And then the old people would warn the children that the seeds are rich and to not eat too many or they will get a belly ache! Séliš-Ql̓ispé culture has always been characterized by close, careful, quiet observation of the natural world, and so our elders would say that only smõmõe, grizzly bears, can eat a lot of the oily nuts at one time, when they are fattening up for hibernation. And only grizzlies have jaws strong enough to crush open the dense cones.

The Séliš and Ql̓ispé have also always known the bird called sn̓ałqʷ, the one known in English as Clark's nutcracker, and as *Nucifraga columbiana* in scientific terminology. This bird depends on sč̓iłpálqʷ and sč̓iłpálqʷ also depends on sn̓ałqʷ. Among all the birds, only sn̓ałqʷ has a beak perfectly shaped to pry open the whitebark cones and extract the seeds. As scientists now also know, sn̓ałqʷ carries the seeds around and caches them here and there across many miles, retrieving many, but leaving some to germinate. The seeds left behind then grow, and in this way sn̓ałqʷ serves as the only way by which sč̓iłp̓álqʷ spreads across the mountains.

The elders tell us scientific methods and technologies can be good too, but only as long as we are careful to use the new ways in service to (under the guidance and direction of) the old ways of knowing, seeing and understanding. The elders have long encouraged us to master and employ the new scientific tools and approaches, but they have also warned that we must never lose the foundation of the traditional way of life: a spiritual and material relationship of respect with sč̓iłpálqʷ, and all the other plants and animals, and the lands, and the waters.

Clark's Nutcracker (Nucifraga columbiana), Rocky Mountain National Park, Colorado. Photo from Wiki Commons, 29868513916, by Donald Hobern.

Introduction to the Confederated Salish and Kootenai Tribes and Their Lands

The CSKT include the Salish, Kootenai and Upper Kalispel Tribes, who originally lived and traded throughout the northwestern region of the lands now known as the United States into the southern part of the lands now known as Canada. Their homelands include what are now portions of the states of Montana, Idaho and Oregon, and range into the Columbia River Basin in Washington State (CSKT – CRP 1996a). The tribes ceded most of this land to the U.S. government on July 16, 1855, through the Treaty of Hell Gate, reserving the right to hunt, fish, graze and harvest in open and unclaimed lands within the tribes' aboriginal territories (Devlin 2021; Bigart and Woodcock 1996); and also reserving from cession, as unceded sovereign land, certain areas including the Flathead Reservation in northwest Montana. The reservation is located in the lower quarter of the Flathead River Basin, including the southern half of Flathead Lake and the Lower Flathead River Basin, including the southern half of Čłq̓étkʷ (Flathead Lake) and Ntx̣ʷétkʷ (the Lower Flathead River). As of 2021, the Flathead Reservation is 1.317 million acres (532,971 ha) in size, of which 815,669 acres (330,090 ha) are owned by the tribes as a whole, and 29,956 acres (12,123 ha) are owned in trust by Native American individuals (CSKT 2021).

As the first to organize a tribal government with a constitution and a corporate charter under the Indian Reorganization Act of 1934, the CSKT are governed by a ten-person Tribal council (CSKT 2021). In May 2020, there were 8,059 enrolled tribal members with 5,360 living on the reservation (CSKT – Annual Reports 2021). The total population of the reservation was 31,394 in 2019 with American Indians accounting for approximately 32% of the entire reservation population (U.S. Census Bureau 2012; CSKT – Annual Reports 2021).

The tribes' mission is guided by traditional principles and values, which form the basis for all facets of tribal operations and services. Critical components include investing in their communities to ensure their ability to function as a completely self-sufficient society and economy, while simultaneously striving to provide sound environmental stewardship that preserves, perpetuates, protects and enhances natural resources and ecosystems (CSKT – Annual Reports 2011). CSKT's vision is to maintain traditional principles and values in the ways they govern their reservation and represent themselves to the rest of the world; and in the ways they continue to preserve their right to determine their own destiny (CSKT – Annual Reports 2011). Today, the CSKT are recognized as a model of a self-sufficient sovereign indigenous nation within the United States. Their tribal government offers a number of services to tribal members and is the chief employer on the reservation (CSKT 2021).

Climate Impacts to Montana Including the Flathead Reservation

> As the leaders of the world's nations have failed to effectively reduce the output of CO_2 and halt the emissions of other harmful industrial by-products, we Indigenous peoples find ourselves as leaders in the movement to prepare for a changing climate, and even more importantly, leading by following the ways and values of our ancestors, demonstrating an alternative path from the destructive ways of colonialism and capitalism. [CSKT Climate Change Strategic Plan, September 2013, updated April 2016 (CSKT – Climate 2016)]

CSKT has compiled detailed information of current and projected climate impacts to Montana and the Flathead Reservation, as part of their 2013 Climate Change Strategic Plan (updated in 2016) and as part of the new Climate Adaptation Plan scheduled for release in 2021 (CSKT – Climate 2016). Climate data and projected impacts to the Flathead Reservation are available through collaborating institutions. The Montana Climate Assessment (MCA), completed in 2017, assessed current and projected future climate change for all of Montana, including the northwestern region that encompasses the Flathead Reservation (MCA 2017; Whitlock et al. 2017). The Montana Climate Office (MCO) provides high-quality climate information and services to all Montanans (MCO 2021). According to the MCA, annual average temperatures across the state have risen between 1950 and 2015 by 2.7°F (1.5°C), with winter and spring temperatures rising by 3.0°F (1.7°C) over the same time period. On the Flathead Reservation, the total temperature increase has been 2.5°F (1.4°C), while the number of days above 90°F (32°C) per year has tripled in the last century (Whitlock et al. 2017).

Projections for average annual warming in Montana range from 4.5°F to 6.0°F (2.5–3.3°C) by mid-century (2050), depending on the greenhouse gas (GHG) emissions scenario.

The lower range results from the moderating (RCP 4.5) scenario in which atmospheric CO_2 concentrations are stabilized by mid-century (through active mitigation measures) and the higher range is from the accelerating (RCP 8.5) scenario, which assumes emissions continue to rise on a "business as usual" path. By the end of the century, average annual temperatures in Montana are projected to increase by 5.6°F to 12.9°F (3.2–7.2°C), depending on the emissions scenario. These changes are expected to be consistent throughout the state, with the most significant temperature rises occurring in midsummer (Whitlock et al. 2017).

CSKT research partners from the Native Waters on Arid Lands (NWAL) program (McCarthy 2020) and the Montana Climate Office produced downscaled climate projections for the Flathead Reservation and extracted relevant data related to crop, livestock and forestry agriculture to inform the 2021 Climate Adaptation Plan (CSKT – Climate 2016; NWAL 2020; MCO 2021). These data are available for two elevations on the Flathead Reservation – the lower elevation (3,065 ft or 934 m) represents the Mission Valley where crops are grown, livestock are grazed and where the human population lives. The higher elevation (6,130 ft or 1,868 m) corresponds to a midpoint in the Mission Mountains, a range of the Northern Rockies located east of the valley, where snowpack accumulates and CSKT manages the forests of the Mission Wilderness. The highest peak in the Mission Range is 9,820 ft (2,994 m). Average annual historical temperatures for the period 1950–2000 in the Mission Valley were around 46°F (7.8°C); similarly, in Mission Mountains, average temperatures were around 32°F (0°C) at the midpoint. Average annual precipitation during this period was around 12 inches (30 cm) in the valley and approximately 60 inches (152 cm) in the mountains, with the latter falling primarily as snow.

Extracted climate projections show annual average temperatures to be 6–12°F (3–7°C) warmer than the historical average (depending on the emissions scenario) in both locations (NWAL 2020). Observations from the accelerating emissions scenario (the path the world is currently on) include the following: (1) average annual precipitation amounts are expected to stay near historic levels in the valley and increase around 5% in the mountains, with summer precipitation totals decreasing slightly and winter levels increasing, accordingly; (2) the frequency and intensity of extreme storms are expected to increase, especially in the mountains; (3) by 2100, the first snowfall will arrive about five weeks later than it has historically; (4) significant warming temperatures in the mountains – with frost-free days doubling from 75 to over 150 annually – will result in more rain and less snow during the winter, leading to significant declines in spring runoff; (5) growing degree days in the valley (defined as the first of at least six consecutive days with average temperatures over 50°F (10°C) will more than double by 2100; (6) the average number of hot days with temperatures at least 100°F (38°C) reaches over 30 in the valley by 2100, from a baseline close to zero, historically.

Overall climate change in Montana is leading to longer growing seasons with hotter temperatures and less spring runoff, which is expected to reduce crop yield for grass hay, alfalfa and barley and degrade the quality of rangeland forage for livestock (Whitlock et al. 2017; Zhao et al. 2017; Polley et al. 2017). Warmer stream temperatures may adversely impact fish populations and spawning rates (USGS 2021). Increased temperatures (especially warmer nighttime temperatures) stress native vegetation, allowing invasive plants to increase on rangelands, and allow crop and forest pests and pathogens to thrive (Finch 2012). One of the most impactful results of warming temperatures and reduced water supplies is increased risk of catastrophic wildfires. Warming temperatures have caused widespread changes in fire regime vegetation compositions, structures, functions and area extent, with increased wildland fire behavior (CSKT – Climate 2016; Barbero et al. 2015; Abatzoglou and Williams 2016; Coates et al. 2016). Wildfire risks and climate impacts are location dependent and occur at various scales, both temporally and spatially. Impacts in the high elevations in the Mission Mountains (home to the whitebark pine) differ significantly from those in the dry grass and sage steppe ecoregions of the Mission Valley, including areas where native vegetation has been replaced with introduced pasture grasses or planted crops. The combination of warming temperatures, reduced snowpack and changes in wildfire regimes is already impacting the health and survivability of the whitebark pine (Fryer 2002; Keane and Arno 1993; Keane et al. 2012).

CSKT Worldview as a Foundation for Developing Tribal Responses to Climate Change

The Séliš, Ql̓ispé and Ksanka people are the keepers of thousands of years of place-based traditional knowledge. Their relationship with the environment was, and continues to be, rooted in culture, tradition, experience and practice. Native stories and songs shared through generations provide evidence of significant ecological events that occurred as humans emerged onto a land already populated with plants, animals, rocks, soils and minerals. As in many other Indigenous communities, Traditional Ecological Knowledge (TEK) has guided climate adaptation among the Séliš, Ql̓ispé and Ksanka people for millennia and continues to underpin climate resilience today. The elders often remind us that it is now our responsibility to care for those who cannot care for themselves, including all of our human and non-human relations.

Within this expansive worldview, each of the three tribes of the Flathead Reservation, Séliš (Salish or "Flathead"), Ql̓ispé (upper Kalispel or "Pend d'Oreille") and Ksanka (Kootenai), is culturally unique and has its own belief system, while remaining similar in certain ways; fundamentally, each tribe holds traditional knowledge of the natural world, and each has profound respect for all of creation. Both traits have enabled the tribes to survive for thousands of years (CSKT – CRP 1996a), using subsistence patterns developed over generations of observation, experimentation and spiritual interaction with the natural world, which has in turn created a body of knowledge about the environment closely tied to seasons, locations and biology. This way of life is suffused with rich oral history and a spiritual tradition in which people respect the animals, plants and other elements of the natural environment. Through the teachings of elders these tribal ways of life continue to this day (CSKT Cultural Preservation Office 2000).

Core tribal values are imbued with long-held cultural attitudes toward the land and its historical use of resources by

Tribal elders sharing knowledge and stories with CSKT children; from left to right: Bud Barnaby, Mike Durglo Sr. and Pat Pierre. Photo credit: Bernie Azure.

ancestors and are based on the fundamental belief that humans and nature are one. They include (1) learning to live in harmony with each other and with the land; (2) understanding that our ancestors left this land to us and that we are only borrowing it from our children. For this reason, as we make decisions that affect the land, we must consider the consequences those decisions will have for the next seven generations, at least; (3) respecting the land, replacing what we take and taking only what we need – by respecting the land, we respect ourselves; (4) preserving spiritual and cultural practices and resources, acting on a spiritual basis when dealing with the environment, always being aware of and appreciating what we have and turning to the Creator when in need; (5) protecting Mother Earth and repaying the land for its gifts; (6) preserving an abundance of native animals, plants and fish; (7) maintaining traditional practices of gathering, hunting and fishing.

In deciding how best to respond to the existential threat of climate change, CSKT planning staff considered two questions that flowed directly from shared tribal values: "If you were given the opportunity to visit the Flathead Reservation 100 years from now, what should the quality of life be like?" And, "What natural and human resources should the reservation produce a century from now, and in what condition should the environment be?" The responses to these questions reflect cultural and traditional ties to the region extending back thousands of years

To convey the integral nature of humanity and the natural and built worlds of the Salish, Kalispel and Kootenai peoples, the shared values of the three tribes have been formalized into a "Lifeways Wheel." The wheel illustrates a tribal perspective for climate adaptation captured in the CSKT Climate Strategic Plan and the CSKT Climate Adaptation Plan (CSKT – Climate 2016). Within the Lifeways Wheel, the circles of life and relationships radiate from the individual and family at the center outward to community and culture, tribe, nation and the world. The spokes on the Lifeways Wheel are the pillars of CSKT

A. Individual / Family **B.** Community / Culture **C.** Tribe **D.** Nation **E.** World

The Confederated Salish and Kootenai Tribes Lifeways Wheel illustrating their worldview as a foundation for responding to climate change. Illustration credit: CSKT.

governance and represent tribal departments that have a key role in adapting to a changing climate: Forestry, Wildlife, Fish, Water, Infrastructure, People, Lands and Air. Within the Lifeways Wheel, all of the departments are woven together with braided sweetgrass, symbolizing healing, peace and tradition. The sector illustrations on the Lifeways Wheel highlight elements of tribal traditional life impacted by a warming world and changing climate (CSKT – Climate 2016), including forests populated by whitebark pine and huckleberries, wildlife relations, moose and bear, flowing streams home to bull trout and other fish, water flowing over the Séliš Ksanka Ql̓ispé Dam (formerly Kerr Dam), lifesaving infrastructure corridors over the main highway through the Reservation, our people (elders, youth and community members) honoring the river, the fertile lands of Snyel̓mn Čłčewm (the Mission Valley) and the sweet air of our homelands.

As with all things, the people of the Confederated Salish and Kootenai Tribes follow the circle of life. The elders always say that all things are connected and that any impact or action on one resource will impact all. To better reflect this perspective, the CSKT Climate Change Advisory Committee (CCAC) analyzes impacts of climate disturbance on tribal resources as a whole. Over the last 50 years, the creation of tribal departments to manage natural and cultural resources has been shaped by Western (colonial settler) views of science, resource management and bureaucracy. Until recently, this influence has been an unavoidable part of tribal self-governance, but this is beginning to shift. This shift is being driven by the scope and extent of climate change impacts permeating and connecting all elements of life. Western reductionist approaches to managing people and the environment as separate sectors become ineffective in a living, breathing, interdependent system. Climate change affects all aspects of our shared environment at once. We are all in this together and we can only adapt, together. Climate change is increasingly seen as a problem of unintended consequences, which emerges when we fail to recognize that all things are connected.

The lesson of interconnection has been a part of traditional ecological knowledge for thousands of years. Climate change is not just an environmental problem resulting from increased atmospheric carbon dioxide in the atmosphere, but also a problem that extends from the denial that individual actions have a collective force and effect. These collective effects defy individual attempts to control and confine them, and so they demand from us our collective recognition and responsibility in order to manage them (Mitchell and Kwasset 2020). To that end, the CSKT CCAC is identifying climate resilience actions that transcend the department-by-department approach, drawing more from traditional knowledge and the idea of restoration and reciprocity in stewarding Mother Earth in a warming climate (Kimmerer 2011; Kimmerer 2013). CSKT climate adaptation approach is guided by the spirit of il̓aw'yeʔ (Eel-lah-wee-yeh), meaning great-great-grandparent, the whitebark pine tree who watched over the people and shared her wisdom for over a thousand years.

CSKT Climate Change Advisory Committee Chairman, Mike Durglo Jr., with il̀awy̓eʔ (Eel-lah-wee-yeh) or "great-great-grandparent"; an ancient whitebark pine that was estimated to be more than 1,000 years old when still alive. Photo credit: Mike Durglo Jr.

Sc̓iłpálqʷ seedlings, Junctiion Mountain, Bob Marshall Wilderness, 2011. Photo by Thompson Smith.

Mature sc̓iłpálqʷ (whitebark pine) with cones, Northern Bitterroot Range, Montana-Idaho border, 2010. Photo by Thompson Smith.

Mature sc̓iłpálqʷ (whitebark pine) with cones, Tex Creek, Beaverhead Mountains, Montana, 2013. Photo by Thompson Smith.

The CSKT Strategic Climate Plan as a Beginning

In recognition of the existential threat posed by a warming climate, CSKT completed their first-ever Strategic Climate Plan for the Flathead Reservation in 2013 (CSKT – Climate 2016). The planning process employed a "Collective Impacts" model to better understand climate threats and galvanize adaptation efforts across tribal departments and programs. Through a series of meetings and workshops organized by the CCAC, tribal stakeholders, elders and participating subject matter experts identified management actions to address collective impacts and develop monitoring and evaluation metrics to assess progress toward creating the environment and quality of life envisioned for their children and their children 100 years in the future (CSKT – Climate 2016).

Tribal elders have recognized, prepared for and responded to changes in climate for millennia. It must be acknowledged that tribal adaptation to climate change as it has occurred over that period of time has been in response to natural variability, which was generally much slower and less extreme than what we are beginning to experience now, and what is predicted in coming decades. Nevertheless, the cultural knowledge and long experience of the Séliš, Ql̓ispé and Ksanka provides a foundation for us as we move into an uncertain future. As such, the CSKT Climate Change Strategic and Adaptation Plans (CSKT – Climate 2016) draw upon that foundation in developing effective, culturally sensitive climate adaptation and mitigation strategies needed to ensure healthy human and ecological communities on the Flathead Reservation for generations to come. Within five years of completing the original climate strategy, tribal members recognized that climate change was unfolding far more quickly than originally projected by scientific models. In 2018, the Department of Historical Preservation – in collaboration with the tribes' leadership, administration, elders, scientific leaders, community members, regional experts and other stakeholders – initiated a process to revise and update CSKT's Strategic Climate Plan and to identify ongoing and future adaptation actions. Historical information was adapted from the Flathead Reservation Comprehensive Resources Plan (CSKT – CRP 1996a, 1996b) and local climate change scenarios were adapted from Native Waters on Arid Lands (McCarthy 2020). Traditional Ecological Knowledge was provided by the Séliš-Ql̓ispé Culture Committee, Kootenai Culture Committee and Historic Preservation/Cultural Preservation Department (CSKT – Climate 2016). Local impact assessments for several sectors (forestry, land, water, fish, air, water, wildlife, infrastructure, people and culture) were developed by CSKT tribal departments and partner organizations. Traditional knowledge, elder interviews, community survey results and supporting photographs and data underpin a circle of life approach to climate resilience (CSKT – Climate 2016). As with all other things, the new CSKT Climate Adaptation Plan is a "living document" designed to guide tribal resilience decisions and actions to conserve, restore and protect forests, land, water, air and all human and non-human inhabitants for generations to come. Restoring and protecting whitebark pine is an example of a climate adaptation action connecting the lifeways of forests, wildlife, water and people.

Whitebark Pine Forests as an Early Focus for CSKT Climate Adaptation

Beginning in 2013, the Flathead Reservation's ancient forests of sčiłpálqʷ (whitebark pine) became an early focus for climate adaptation actions by CSKT. Whitebark pine is a gymnosperm, and a member of the division Coniferophyta (USDA – NRCS 2019). It is the only North American species of the "stone pine" group and can be a shade-tolerant, slow-growing, long-lived tree (Minore 1979). Mature whitebark pine trees will age to over 400 years, especially on harsh growing sites and older individuals in excess of 1,000 years have been found (Luckman et al. 1984; Perkins and Swetnam 1996). Whitebark pine is slow growing in both height and diameter and, although early to establish post-disturbance, will often be surrounded or topped out by faster growing species from the same establishment cohort (Arno and Hoff 1990).

Whitebark pine has the largest distribution of all five-needle white pines in North America (Tomback and Achuff 2010; Little and Critchfield 1969). It is found in the subalpine and treeline forests of the United States and Canada, including the northern Rocky Mountains, Great Basin, Sierra Nevada and Cascades and northern coastal ranges (Little and Critchfield 1969; Arno and Hoff 1990; McCaughey and Schmidt 2001). Whitebark pine forms extensive forests in the northern Rocky Mountains of the United States and it is also abundant on the eastern slope of the Cascades and Coast Ranges; mosaic-like stands tend to form at the northern end of its distribution (Arno and Hoff 1990). On the Flathead Reservation, whitebark pine is found generally on all slopes and aspects between 1,800 and 2,800 m elevation (5,900–9,200 ft); it will rarely establish either above or below this elevation band. Whitebark pine grows on poorly developed cold soils, and in deeper volcanic ash deposits (USDA 1975) and only starts cone production at 30–60 years of age. Cone production maximum is not attained until reaching a mid-upper canopy or open grown stature, around

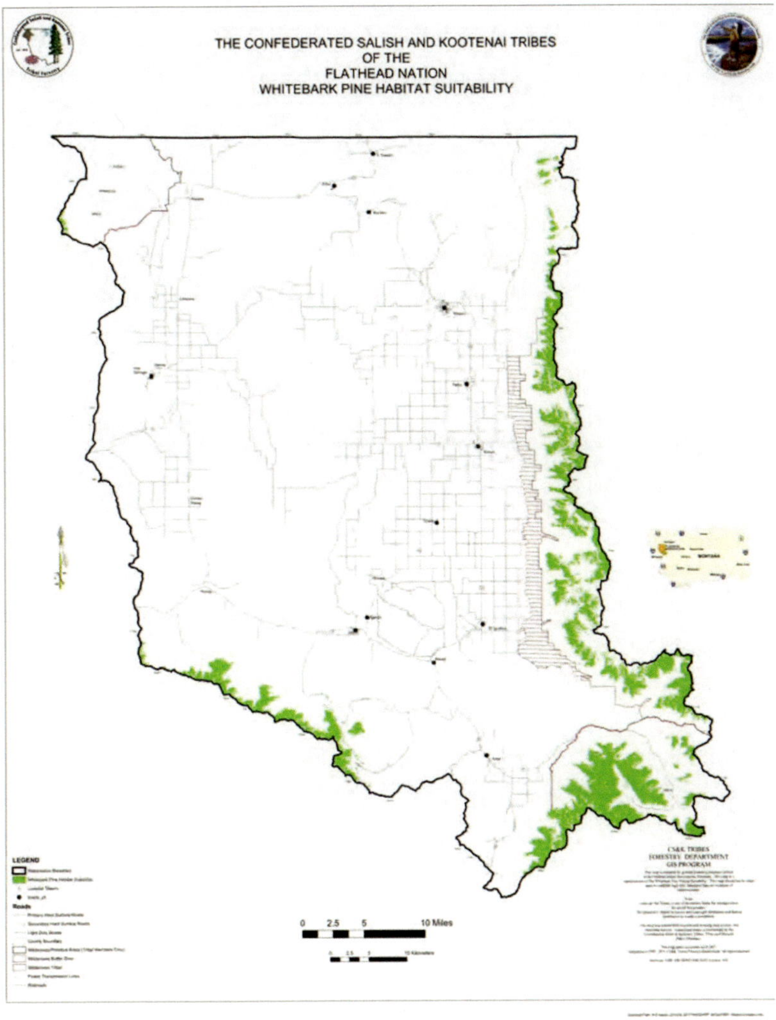

Map of Flathead Reservation depicting areas of suitable whitebark pine habitat. Map credit: Confederated Salish and Kootenai Tribes.

125–250 years of age (Arno and Hoff 1990). The frequency of large, cone crops within a given stand varies regionally by climate and population level and ranges from yearly to every two or three to five years in many areas, with a few cones typically produced within a stand every year (Crone et al. 2011; Tomback et al. 2001; Arno and Hoff 1990).

Whitebark pine forests occur in two high-mountain biophysical settings, about which the Séliš, Qlispé and Ksanka people hold an intimate knowledge. On productive upper subalpine sites, whitebark pine is the major seral species that is replaced by the more shade-tolerant maninłp (subalpine fir, *Abies lasiocarpa* [Hook.] Nutt.), tstsélp (Engelmann spruce, *Picea engelmannii* Parry ex. Engelm) or płtin'éʔ (mountain hemlock, *Tsuga mertensiana* [Bong.] Carrière), depending on geographic region (Arno and Weaver 1990). These sites support upright, closed-canopy forests in the upper subalpine lower transition to timberline, just above or overlapping with the elevational limit of the shade-intolerant qʷqʷliʔt (lodgepole pine, *Pinus contorta* Douglas ex. Louden) (Arno and Weaver 1990; Pfister et al. 1977), allowing the two species to be codominant. Other forest codominants found with whitebark pine on these sites are cq̓ełp (Douglas-fir, *Pseudotsuga menziesii* [Mirb.] Franco), k̓ʷxʷtné (limber pine, *Pinus flexilis* James) and čtx̣ʷey cáqʷlš (alpine larch, *Larix lyall Parl*) (Pfister et al. 1977). Fire regimes on these sites tend to vary from moderate-frequency/moderate-effects to low-frequency stand replacement events.

Whitebark pine can also successfully dominate high-elevation settings on harsh sites in the upper subalpine forests and at treeline on dry, high-elevation slopes (Arno 1986; Arno and Weaver 1990). Other species, such as subalpine fir, spruce and lodgepole pine, can occur on these sites as infrequent individuals with poor growth (Arno and Hoff 1990; Arno and Weaver 1990; Pfister et al. 1977). Alpine larch is often found on north-facing mature stand whitebark pine sites (Arno and

Habeck 1972). Whitebark pine can also form krummholz and other typical high-elevation growth forms and stands in the alpine treeline ecotone (Arno and Hoff 1990; Tomback 1989) and as a minor component in lower subalpine sites (Cooper et al. 1991; Pfister et al. 1977). As tribal elders have long noted, with the open-grown nature of these stands, fires tend to be infrequent, low intensity and relatively small in size.

Whitebark pine forests form a wide variety of higher-elevation plant communities (Tomback and Kendall 2001). Most whitebark pine forests, especially the seral types, have relatively low diversity in vascular plants (Forcella 1978), with the majority of understory plant cover composed of ssipt (grouse whortleberry, *Vaccinium scoparium* Leiberg ex. Coville), stša (huckleberry, *Vaccinium membrenaceum* Douglas ex. Torr.), menziesia (*Menziesia ferruginea* Sm.), Hitchcock's smooth woodrush (*Luzula glabrata* [Hoppe ex. Rostk.] Desv. var. hitchcockii [Hämet-Ahti] Dorn) and sl̓čest'yeʔ (common beargrass, *Xerophyllum tenax* [Pursh.] Nutt), with sedge (*Carex rossii* Boott) and Geyer's sedge (*C. geyeri* Boott), Pink Mountain-heath (P*hyllodoce empertriformis* [Sm.] D. Don) and broadleaf arnica (*Arnica latifolia* Bong) – depending on geographical area, aspect and elevation – also form part of the understory vegetative component (Keane and Parsons 2010; Pfister et al. 1977). Other plants that may be occasional dominants include skʷlsé(łp) (kinnikinnick, *Arctostaphylos uva-ursi* [L.] Spreng.) Idaho fescue (*Festuca idahoensis* Elmer), Parry's rush (J*uncus parryi* Engelm.) and Wheeler bluegrass (*Poa nervosa* [Hook.] Vasey) (Arno and Weaver 1990; Aubry et al. 2008). High-elevation climax stands of whitebark pine support many unique alpine, subalpine and montane undergrowth species assemblages, some of which are only found in association with whitebark pine (Tomback and Kendall 2001). An early study (Forcella and Weaver 1977) found that whitebark pine forests had unexpectedly high biomass, but low productivity.

Ecological Interactions and Importance

The whitebark pine is an important source of food for seed-eating birds and mammals (Keane et al. 2017). The high nutritional value in whitebark pine seeds is characterized by 21% protein (Lanner and Gilbert 1994) and fat content ranging from 28% (Robbins et al. 2006) to 52% (Lanner and Gilbert 1994). The seeds are large, weighing an average 0.175 g each, with thick seed coats (McCaughey and Tomback 2001; Tomback and Linhart 1990). This rich, tasty seed attracts sn̓ałqʷ (Clark's nutcracker, *Nucifraga columbiana*), a major seed disperser of the pine. Clark's nutcrackers each cache about 30,000–100,000 seeds a year in small, widely scattered caches, usually under 2–3 cm (0.75–1.25 in) of soil or gravelly substrate. Nutcrackers retrieve these seed caches during times of food scarcity and to feed their young. Cache sites selected by nutcrackers are often favorable for germination of seeds and survival of seedlings. Those caches not retrieved by the time the snow melts contribute to forest regeneration. Consequently, whitebark pine often grows in clumps of multiple trees, germinating from a single cache of two or more seeds (Tomback et al. 2001).

Up to 120 other species are known to utilize the whitebark pine (Keane et al. 2012). ʔiscč̀ (pine squirrel or red squirrel, *Tamiasciurus hudsonicus*) cut down and store whitebark pine cones in their middens. Smxẹyčn (Grizzly bear, *Ursus arctos*) and ṇłámqey'n (American black bear, *Ursus americanus*) often raid squirrel middens for whitebark pine seeds, an important pre-hibernation food. Squirrels, kʷlkʷlé (northern flicker, *Colaptes auratus*) and n̓łqʷeyqʷay̓ á (mountain bluebird, *Sialia currucoides*) often cavity nest in whitebark pines, and bear, tšeć or snečłće̓ʔ (elk, *Cervus canadensis*) and other vertebrates use mid-to-high-elevation whitebark pine communities as summer habitat.

Ecological Concerns

Whitebark pine forests are currently in decline throughout their range in North America because of the combined effects of historical and current fire exclusion management policies, mountain pine beetle (*Dendroctonus ponderosae* Hopkins) outbreaks since the 1980s and the invasive pathogen *Cronartium ribicola*, which causes the disease white pine blister rust in five-needle white pines (Keane and Arno 1993; Kendall and Keane 2001; Tomback and Achuff 2010). The loss of this high-elevation tree species poses serious consequences for upper subalpine ecosystems, both in terms of the impacts on biodiversity and in losses of both biotic and abiotic ecosystem processes and services (Tomback et al. 2001; Tomback and Achuff 2010).

White Pine Blister Rust

Most stands of whitebark pine across the species' entire natural range are infected with an introduced fungal pathogen, white pine blister rust (*Cronartium ribicola*). In the northern Rocky Mountains of the United States, whitebark pine mortality in some areas exceeds 90% (Keane et al. 2012). The blister rust also has overwhelmed commercially valuable western white pine in these areas and infected k̓ʷxʷtné (limber pine, *Pinus flexilis*) populations as well (USDA – USFS 2021). There is currently no effective method for controlling the vectoring and disease effects of blister rust (USDA – USFS 2021; Keane et al. 2012). A single rust infection of a pine might establish and then spread for years. Once inside the pine needle, the fungus grows into the twigs and ultimately to the main stem of the tree. The damage is caused by the rust killing the cambium, causing a canker, ultimately girdling the stem which prevents water and nutrients from passing through the canker area; as a result, the twig, branch or stem dies. When the canker forms on the main stem, it causes top kill, often resulting in deformation or death of the tree. During the course of the disease, branch death and top kill significantly reduces cone production and tree vigor. A small number of trees (fewer than 5%) in most populations seem to possess genetic resistance to blister rust (Sniezko et al. 2018). Restoration efforts undertaken by forestry practitioners, tribal, federal government and private, in the northern Rocky Mountains involve harvesting cones from resistant whitebark pines, growing seedlings and planting seedlings in suitable sites.

Mountain pine beetle (*Dendroctonus ponderosae*) is a species of bark beetle native to western North America that may have also contributed to the widespread destruction of whitebark pine stands. Beetles lay their eggs and introduce pathogenic fungi into host trees, infecting many species of pine (Keane et al. 2012). This combination of larval development and fungal colonization is characteristically adequate to kill old or unhealthy trees. However, because of increased climate warming, the beetles have expanded their attacks to include younger, healthier trees (Kichas et al. 2020). Since the 1980s, the climate at high elevations has increased temperature enough for the beetles to reproduce within whitebark pine forest communities, often completing their life cycle within one year and enabling their populations to grow exponentially. Entire forest landscapes in the northern Rocky Mountains have become expanses of dead gray whitebark skeletons (Six and Bracewell 2015). In 2007, for example, the U.S. Fish and Wildlife Service estimated that beetles had killed whitebark pines across 500,000 acres (200,000 ha) in the west; while in 2009, beetles were estimated to have killed trees on 800,000 acres (320,000 ha), the most since record-keeping began (Tomback 2018).

Fire Suppression

Incorporation of institutionally aggressive fire suppression has led to slow whitebark pine population declines during both the 20th and 21st centuries by interrupting the disturbance dynamics of stands without the occurrence of low- and moderate-intensity fire effects providing for forest resilience and health, and suppression of high-density stand insect-disease threats (Fryer 2002). In the absence of low-intensity periodic wildfire, whitebark pines in these stands are replaced by more shade-tolerant, fire-intolerant species such as maninłp (subalpine fir, *Abies lasiocarpa*) and t̓st̓séłp (Engelmann spruce, *Picea engelmannii*) (Fryer 2002; Keane and Arno 1993; Keane et al. 2012). This replacement often will lead to the development of a stand-replacement lethal fire regime, maximally perturbing whitebark forests. In addition, senescent and blister rust-infected pine trees are not destroyed by natural periodic ground fires, further diminishing the whitebark pine forest's vitality and survival (Keane et al. 2012; Kichas et al. 2021).

Rapid post-disturbance seed dispersal by Clark's nutcrackers combined with hardy seedling growth results in early whitebark pine community development after fire and other forest community disturbances (Tomback and Kendall 2001; Tomback et al. 2001). Furthermore, whitebark pine seedlings thrive in harsh, high-elevation droughty sites and may eventually act as early- to mid-seral nurse trees to more shade-tolerant conifers and vegetation (Tomback et al. 2001). At upper subalpine elevations, where it is common in many regions (Resler and Tomback 2008), whitebark pine helps regulate snow intercept and subsequent melt and reduce soil erosion (Farnes 1990). Whitebark pine is considered both a keystone species for encouraging community diversity and a foundation species for promoting ecological community stability (Ellison et al. 2005; Tomback et al. 2001; Tomback and Achuff 2010) and as an important ecosystem services provider. The loss of whitebark pine would potentially affect fire regimes, snowpack interceptions and local hydrology and other visual and recreational ecosystem services (Keane et al. 2012; Keane et al. 2017).

The Indigenized Concept of "Biocultural Restoration"

In considering how best to restore and sustain whitebark pine forests across the Flathead Reservation in the face of a shifting climate and other threats, the CSKT are adopting an Indigenized concept of restoration called "biocultural restoration." Biocultural restoration is the science and practice of restoring not only ecosystems but also human and cultural relationships to place, so that cultures are strengthened and revitalized along with the lands to which they are inextricably linked (Kimmerer et al. 2021; Greenlaw et al. 2009). CSKT has undertaken other projects based on this culturally informed approach, including the reestablishment of extirpated species (e.g., trumpeter swans) (Becker and Lichtenberg 2007), and a major restoration effort for the Jocko River (Explore the River) (CSKT Fish-Wildlife 2021). According to this concept, sč̓iłp̓álqʷ isn't just another tree. It is not a mute mass of wood and pulp, but a silent teacher of some of life's most valuable lessons that have been passed by the great-great-grandparent tree, il̓aw'iyeʔ, and many others in the forest.

Seedlings start their lives in high elevations where they are exposed to extreme variation in temperature, high winds and dense snow. The snowpack is so deep that sč̓iłpálqʷ evolved pliant branches able to bend under heavy snow without breaking. Sč̓iłpálqʷ spend several decades, 30–60 years, without ever bearing a single cone. After all that time spent growing in such a harsh environment, it might seem logical to presume that sč̓iłpálqʷ would be a very stingy organism, eking out a meager existence, and offering little to its fellow creatures. But sč̓iłpálqʷ proves otherwise. Full of rich fats and proteins, its pinion seeds provide food for dozens of animal species in the high-alpine forest. Of these animals, the tree has forged special relationships with sn̓ałqʷ, the Clark's nutcracker, and with iscč, the pine squirrel. It might seem that these animals are greedy for taking so many of the pinions and hoarding them away underground; but this stockpiling behavior is balanced by the high number of caches that remain for other animals, like smọ̃mõe, grizzly bears, to dig up and eat. This greediness is also balanced by those occasions when the caching of seed sprouts into a seedling so that sčiłpálqʷ can once again emerge.

Throughout time, the tribes have joined sn̓ałqʷ, and iscč, to collect the seeds, and like the animals, their greediness is balanced. Natural resource workers collect the seeds and send them to the U.S. Forest Service Laboratory in Coeur d'Alene, Idaho, where disease-resistant individuals are identified. Collaborative teams from the U.S. Forest Service, Bureau of Land Management and CSKT later plant these healthy seedlings in high-elevation forests and burn scars from recent wildfires. The teams planted over 1,000 seedlings in 2019, and those that survive will outlive their planters by centuries.

These relationships demonstrate the ways that sčiłṗálqʷ embodies humility, patience, perseverance, flexibility, generosity and reciprocity throughout its life cycle, and these lessons resonate in the culture of the Salish and Kootenai people. Despite these lessons, sčiłpálqʷ had not been a major priority of the tribes until climate planners began holding community gatherings, workshops and planning sessiions deliberately focused on combining culural practices and natuiral resource management efforts. It was only through community involvement and dialogue recognizing the interconnecting interests of human beings and the biotic ommunity that sčiłpálqʷ reemerged as a species of tremendous importance. Sčiłpálqʷ has provided for the tribal community for many generations, and now the tribes are doing their part to preserve and protect sčiłṗálqʷ so that both lives are prolonged. The tribes are taking its lessons to heart.

This kind of dialogue is not new to Native people and is often very foreign and sometimes inaccessible for other communities. It is our hope that by sharing this story, others may use it as an example of how to build healthier relationships between human cultures and the biotic communities that support them. Sometimes, this seems like an impossible task, but as ancestors have reminded, the tribes in the past, and elders and sc̓iłpálqʷ continue to remind us today, "Never give up. Never give up."

The Practice of Restoring Whitebark Pine Forests and Ecosystems on the Flathead Reservation

Within the 1.3 million acres (526,091 ha) of the Flathead Reservation, the CSKT Forestry Department manages 256,757 acres (103,905 ha) of forest year-round (CSKT – Forestry 2000). The CSKT Forest Management Plan (2000) prioritizes new actions in forest management based on the principles of ecosystem management, all within the framework of traditional long-term forest management practices (CSKT – Forestry 2000). CSKT Forestry strives to manage sustainable forests for the future through the use of state-of-the-art computer models, satellite imagery and spatial analysis with geographic information systems (GIS). Within this focus, the Forest Management Plan states that "ecosystem management uses ecological, cultural, economic, social, and managerial principles to maintain and restore the ecological diversity and integrity of the reservation's surrounding forests" (CSKT – Forestry 2000).

The CSKT Forestry Department's goal is to restore reservation forest habitat to be as similar to pre-European settlement as possible (CSKT – Forestry 2020). This goal considers the tribe's knowledge of different management practices of ecosystem restoration and how they apply to the different habitats. The 80-year goal from the Forest Management Plan is to enhance sustainable health, reestablish culturally traditional fire regimes and help manage and eliminate negative impacts from overstock, insect and disease on forest health (CSKT – Forestry 2000). Each goal seeks to address issues related to years of fire suppression and accelerating climate change, including increasing temperatures and decreasing annual precipitation (per above). Recognizing that balance between all-natural resources creates a sustainable forest, CSKT is committed to incorporating culture and traditional knowledge in all forest management decisions. CSKT Forestry not only seeks approval from the tribal council on proposed forest management directions and projects but also requests the guidance of the

Séliš-Ql̓ispé Culture Committee and the Kootenai Culture Committee and from elders who voice their position on any project regarding the land. CSKT Forestry then selects the most relevant option to achieve ecosystem management goals (CSKT – Forestry 2020).

Climate change has had – and will continue to have – significant consequences for the health and restoration needs of whitebark pine forests throughout its distribution: both by acting as a stressor itself and by exacerbating other threats to this species (per above). Blister rust and mountain pine beetle outbreaks correlate with temperature increases at high elevations where mountain pine beetles have been historically absent; while a decrease in the occurrence of long, bitterly cold winters has created favorable conditions for mountain pine beetle breeding cycles (Bentz et al. 2009; Mahalovich et al. 2006). Over time, more frequent and severe drought events have weakened individual trees and made them more susceptible to disease, with beetles often present in the ecosystem now to finish off the job. Natural fire regimes have been disrupted significantly in the last century, disrupting natural processes that prevented the encroachment of tree species from ecosystems further downslope from competing with whitebark pine and contributing to the spread of more shade-tolerant species (White and CSKT 2010). Hotter temperatures and an increase in the duration of extremely hot days – both products of climate change – are the source of added ecological stress. Altogether, this combination of stressors has created conditions conducive to a stand-replacing, lethal fire outbreak in whitebark pine ecosystems. Consequently, CSKT is implementing key strategies in restoring whitebark pine to help the ecosystem rebalance by building resilience to each threat.

Whitebark pine and fire – both wildfire and the use of controlled burns as a long-term ecosystem management tool by the tribes – are two symbiotic natural resources that have strong cultural ties to the tribes and Forestry's focus and work. Traditionally, tribes used controlled burns to help sustain forest vegetation, and in so doing, helped maintain natural fire regimes across the landscape; with other ignition sources coming from lightning strikes and fires below the mountain running upslope (White and CSKT 2010). As a species, whitebark pine thrives after a fire has moved through an area and the previous forest canopy and competition are eliminated. Today, with whitepine blister rust and mountain pine beetle invasion, CSKT Forestry and CSKT Division of Fire are working together throughout the year to use controlled burns to clean out dead and woody downed material, to manage forest fuel loads and to prepare forest lands for restoration activities. Controlled burns help prepare an area in advance of planting and restoring forest stands through the use of understory or overcast burns; while simultaneously providing both the forest and forest workers a method for countering the effects from insects, disease and lethal fires in the whitebark pine ecosystem.

The Importance and Practice of Restoring Whitebark Pine Forests on the Flathead Reservation

Managing and restoring whitebark pine ecosystems is important to CSKT because of the benefits provided by this keystone species within whitebark pine forests and the lands below these forests. Not only are whitebark pine seeds a first food culturally, they are also important food sources for numerous wildlife species, including the grizzly bear. Ecologically, the loss of whitebark pine forests negatively impacts soil erosion and water quality for forest ecosystems and natural resources below (Keane et al. 2012).

As first step in the restoration process, Salish Kootenai College (SKC) initiated a student research project in 2013 to survey extant forest conditions, including tree age and fire frequencies, to characterize past forest conditions. SKC students also looked for evidence of human impacts to fire disturbance frequencies. Using this information, CSKT Forestry created their first maps and inventory of whitebark pine in 2016, in collaboration with researchers from the U.S. Forest Service, the Bureau of Land Management, and the Hi5 (whitebark pine) Working Group of the Crown Managers Partnership (CMP-Hi5 2019), a coalition of state, federal and provincial agencies working collaboratively with tribal and First Nations on shared natural resource priorities in the Crown of the Continent landscape (CMP 2021). These documents have guided CSKT Forestry in researching and mapping 105,000 acres (42,492 ha) of optimal subalpine habitat on the reservation. Field observers found that there were signs that white pine blister rust was decimating whitebark pine throughout the region; but interestingly, our survey work has consistently documented trees in infected stands that are still healthy and resisting the fungus. The genetics of these resilient trees – known as "plus trees" (Keane et al. 2012) – are vital in the tribes' restoration work, and foster hope for future whitebark pine survival here and elsewhere.

Since 2016, CSKT reforestation specialists have continued to create an inventory of whitebark pine "plus trees." Each effort begins with the identification of locations that need to be scouted; using maps and existing inventories as the basis for on-the-ground searches for mature trees. In 2019, the Crown Managers Partnership (CMP 2021) began a partnership with a local geographic information system (GIS) firm, Map Monsters, to create detailed maps of the reservation's whitebark pine stand conditions. The resulting conservation value map highlighted additional areas across the reservation for expanded work on our inventory (CMP-Hi5 2019). The results of this landscape-scale collaborative effort are then laddering up to a national-scale restoration effort for this species, with the CSKT playing a leadership role at multiple spatial scales.

The next phase of long-term restoration work involves on-the-ground survey work in whitebark pine forest habitats, and, with a CSKT reforestation forester and Bureau of Land Man-

A lone surviving whitebark pine tree in a stand of deceased whitebark pine trees on the Flathead Reservation. Photo credit: ShiNaasha Pete.

Dead sc̓iłpálqʷ (whitebark pine), Junction Mountain, Bob Marshall Wilderness Area, 2011. Photo by Thompson Smith.

Stand of dead sc̓iłpálqʷ (whitebark pine), Junction Mountain, Bob Marshall Wilderness Area, 2011. Photo by Thompson Smith.

agement-certified tree climber(s), navigating rough logging roads by truck or ATVs in an attempt to get as close as possible to the survey areas. This can be very challenging given that most whitebark pine forests occur at high elevations that do not have (or, in the case of the CSKT Mission Wilderness areas, allow) roads. To travel in these areas requires packing in water, food, iPads, radios, tools, climbing gear and cages for young whitebark pine cones and hiking across survey areas searching for mature cone-bearing whitebark pines and potential "plus trees."

During the summer of 2019, CSKT provided the CSKT Forestry with three student interns to accompany a forester to scout whitebark pine habitat. Trekking over 20 miles of terrain in road-less areas during the day and camping at night, the team was able to identify and survey over 20 mature whitebark pine trees. When searching for a mature whitebark pine tree, teams look for trees that are cone-bearing and that appear safe to climb. Data collection then focuses on three different components: (1) basic information about the tree and its location, e.g., coordinates, date and time, diameter-at-breast-height, tree height, live crown ratio, whether it is a cone-bearing tree, which of the three growth forms this tree falls into and whether the tree is part of a whitebark pine clump; (2) disease information such as blister rust cankers, and any indication of flagging, swelling and oozing. If the tree has signs of a mountain pine beetle infestation, the stage of tree kill is recorded along with the estimated year of the last beetle attack recorded. Animal activity from grizzlies, red squirrels or Clark's nutcracker is marked in this area as well; (3) detailed stand information for the selected tree, including stand age, overstory composition of the three most common tree species, understory composition with the same three most common species and habitat type of the stand. If the tree shows no indications of stress, it becomes a "plus tree" and will continue to be monitored over time for growth and any changes indicating distress or disease in the stand.

The role of "plus trees" in the restoration of whitebark pine is essential. Currently, CSKT foresters collect cones from the very tops of each mature tree by caging a subset of the cones early summer before returning to collect them in the fall per methods developed by the U.S. Forest Service, Rocky Mountain Research Station (Keane et al. 2017). Teams construct their own cages – which are designed to prevent seed depredation by squirrels – using duct tape and 0.25-inch (0.63 cm) fence mesh. The safety of CSKT Forestry-certified tree climbers in such remote areas is of paramount importance, with partners from the Bureau of Land Management or the U.S. Forest Service joining tribal staff at the end of each season to help collect cones. Each tree has an assigned number according to the year with a cone and cage count, with cones placed in burlap sacks, tagged with the tree number, date and time, and GPS coordinates.

Back at the CSKT Forestry Department, sacks of cones are stored at room temperature until cone collection for that season has been completed and the cones are dry and ready for processing. When cones are ready to have their seeds extracted, one sack at a time is processed. Cones – and all seeds – are processed by hand. The seeds from all cones in each sack (which originate from one tree only) are sealed in bags and marked with identifying information, before seed bags are weighed, recorded, sealed and stored (CSKT-Forestry 2000).

The germination process for whitebark pine seeds collected from "plus trees" comes next. In 2016, the first seeds collected on the Reservation were sent to the Coeur D'Alene Idaho U.S. Forest Service Nursery. Originating from nine individual trees, these seeds were the first from CSKT lands to undergo testing for resiliency to the white pine blister rust fungus and complete their last cycle of inoculation testing this in 2021: during inoculation, whitebark pine seedlings are sprayed with white pine blister rust spores six to eight times over a two-year period to identify stock that are genetically resistant to the disease (Mahalovich et al. 2006). Since 2016,

Caged whitebark pine cones after collection by CSKT tree climbers in autumn. Photo credit: ShiNaasha Pete.

Whitebark pine seeds that have been removed from whitebark pine cones by CSKT restoration foresters. Photo credit: ShiNaasha Pete.

CSKT Forestry has successfully germinated their own whitebark pine seeds. Seeds are put through a strategic sterilization and scarifying process CSKT developed, which improves the rate of germination (CSKT – Forestry 2000). They are then sown in preparatory amendment soils and take two years for the seedlings to be ready to ship out for contracts with Crown Managers Partnership Whitebark Pine Working Group. The seedlings are stacked ten to a bundle with 50 bundles packed into a box and placed into the cooler until shipment. The seedlings will be distributed to Flathead National Forest, Glacier National Park, Bureau of Land Management, Lolo National Forest and here on the CSKT Reservation.

When planting whitebark pine, typically one hikes to the site after loading up planter bags that are either belted at the hips or slung over the back. When planting the seedlings, the selected locations would be the south-facing aspect of the mountain. To give the seedling some protection from natural elements, the seedlings are planted on the north side of a rock or tree stump. CSKT Forestry had their first planting of whitebark pine in 2019. The Liberty fire had burned over 400 acres in 2017 in the Jocko Primitive area and left perfect conditions on a 9-acre (3.6 ha) opening on a south-facing aspect, where past whitebark pine growth had been recorded previously. Over 1,900 seedlings were planted in June where the day started early and warm, ending with an early summer snow to cover the freshly planted whitebark pine seedlings. The CSKT planting crew departed in a muddy success, feeling accomplished and proud. Today, those seedlings are thriving with a 92% survival rate. CSKT, looking ahead, has 236 acres (106 ha) of whitebark pine restoration plots planned out in different areas in the Boulder Creek drainage on the east Flathead Lake shore mountain range. This area of the reservation has a large inventory of "plus trees" and has been identified as a refugia area for this species.

Confederated Salish and Kootenai Tribes' greenhouses where the Forestry Department germinates whitebark pine seedlings. Photo credit: ShiNaasha Pete 2020.

The first whitebark pine seedling planted in the mountains of the Flathead Reservation after the 2017 Liberty Fire. Photo credit: ShiNaasha Pete.

Moving forward, CSKT Forestry's restoration goal ls for whitebark pine are to continue to scout, increase "plus tree" inventory, stock a whitebark pine seed bank for future fire, harvest restoration sites and create a high-elevation whitebark pine plan- tation; this includes applying proper management practices such as understory and broadcast burns, prepping stands by eliminating competition from non-whitebark pine species, daylighting activities to open the forest canopy and growing supplemental seedlings in CSKT Forestry green houses.

Cultural Connections to Whitebark Pine

Restoring whitebark pine forests across the Flathead Reservation has reconnected the people back to the meaning of this tree in tribal culture, rekindling traditional memories and stories like the ones shared in this chapter. The tribes now have significant opportunities to teach Native youth about the cultural importance of whitebark pine to carry into the future, through new internship and training programs for youth. One such program that has been recognized as exemplary throughout the state is Environmental Advocates for Global and Local Ecological Sustainability, or EAGLES. Awarded grants from the Dreaming Tree (a philanthropic foundation) have helped fund biological field technician internships for local high school and tribal college student internships, given the importance of teaching youth about ecosystem management and sustainable stewardship practices for the future. We are in need of saving our people, culture and land, and it is up to adults today to start teaching the young adults and youth. This is the most imperative goal, not only for forest management, but for whitebark pine as well.

Today, educational outreach is a necessity in the Environmental Science fields; but it is through the strength of CSKT's tribal culture that this keystone species will persevere through the hope, faith and effort that the restoration foresters, interns and students are putting into a species that they will never see thrive in their life-times. Nonetheless, the tribes feel strongly about sharing their efforts as a way of encouraging people to begin recognizing what is being lost today through global change. One way CSKT addressed this was by sharing our cultural connection to nature, and whitebark pine specifically, through a 2021 PBS documentary called "Ghost Forests" that featured work by CSKT on this species on the Flathead Indian Reservation (McCabe 2021). The film allowed viewers to accompany restoration foresters in the CSKT (among others) through the extraordinarily labor-intensive process of restoring whitebark pine forests across its distribution. Ongoing projects include a children's book and a virtual reality video across whitebark pine forests.

Seeing with Both Eyes to Restore and Protect Whitebark Pine

"Sčiłpálqʷ : Biocultural Restoration by the Séliš (Salish) and Ql̓ispé (upper Kalispel) Nations of the Flathead Reservation" illustrates the power of seeing with both eyes, or "two-eyed seeing," to conserve and restore whitebark pine. Two-eyed seeing is rooted in the concept that knowledge derived from Native wisdom and Western science, when woven together, allow the "seer" to create more holistic approaches for protecting our world for future generations (Marshall 2017; Colbourne et al. 2019; MacRitchie 2018; Peltier 2018). Underlying the two-eyed seeing approach are the concepts of humility and respect. No single person holds all knowledge — "humility" reminds us of that fact and "respect" allows different perspectives to be woven together.

The tribes' success story is not that they planted 2,000 trees, it is that they are restoring a lost connection, restoring understanding and restoring respect for the world that nurtures us. The story of the whitebark pine describes its longevity, decline, and now its revitalization mirrors the strength, loss and return of Indigenous ways of knowing and caring for Mother Earth. Indigenous people around the world have lived through major climate shifts before. They have survived, adapted and thrived due to their ability to see, learn and pass on wisdom about how to sustainably steward the forests, land, waters, air and wildlife upon which they depend. Biocultural restoration brings this traditional knowledge forward, and joins it with scientific observations, data and models of the ecological impacts of anthropogenic-driven global warming, to create a sustainable path that can reshape humanity's response to climate change. In turning to the teachings of Indigenous communities, teachings like the concept of reciprocity and respect for the environment, we can transition our path through climate change crisis from destruction to resilience (Mitchell and Kwasset 2020). Tribal communities' capacities for adaptation and restoration, which are rooted in millennia of observing and respecting the world around us, offer the world hope and wisdom.

Fundamental to this path are the relationships of people to place and the concept of reciprocity. In the words of Robin Wall Kimmerer of the Citizen Potawatomi Nation,

> For much of human's time on the planet, before the great delusion, we lived in cultures that understood the covenant of reciprocity, that for the Earth to stay in balance, for the gifts to continue to flow, we must give back in equal measure for what we take. (Kimmerer 2011)

Reciprocity requires we focus on the needs of the natural world as much as we do on our human needs. Every gift received from the natural world requires we give back in equal or greater measure. Restoration becomes an act of love for the world that sustains us and our gift to the seven generations that follow in our footsteps.

CSKT schoolchildren learning about whitebark pine restoration efforts and climate change from members of the CSKT Forestry Department during a field trip to these ancient, high-elevation forests. Photo credit: Anne A. Carlson.

Tony Incashola, Jr. (above) and Mike Durglo, Jr. (below) speaking to University of Montana students and staff about the Confederated Salish and Kootenai Tribes' efforts to restore whitebark pine on the Flathead Reservation. Photo courtesy of Mike Durglo, Jr.

Acknowledgments

The authors would like to express their deep gratitude to: the Tribal Council of the Confederated Salish and Kootenai Tribes; the Bureau of Indian Affairs Tribal Resilience Program for multiple funding awards from 2018 through 2021; the Roundtable of the Crown of the Continent Adaptive Management Initiative through the Kresge Foundation; the Great Northern Landscape Conservation Cooperative; Dreaming Tree; the Wilderness Society; the Crown Managers Partnership (CMP) and members of the CMP Hi5 Working Groups for additional funding and/or partnership; and all members of the CSKT Climate Change Advisory Committee. The CCAC is grateful to Ron Oden (Ron Oden Designs) for his artistic rendering to the Lifeways Wheel. Local climate change scenarios were provided by the Native Waters on Arid Lands Program (USDA/NIFA AFRI 2015-69007-23190) and the Montana Climate Office. The contributors would like to express their special appreciation for the work of our colleague Shirley Trahan, senior translator/transcriber and language advisor for the Séliš-Ql̓ispé Culture Committee. Shirley worked through Salish-language recordings of tribal elders from the 1970s and provided our project with something of the greatest importance: bilingual transcripts of passages relating to sčiłpálqʷ (whitebark pine) and its importance in Séliš-Ql̓ispé culture. *Lemlmtš* (thank you) Shirley. Most of all, we are indebted to the Séliš-Ql̓ispé Culture Committee and Elders Advisory Council. Since the CSKT's reestablishment of the elders group as an organized entity in 1975, the elders have generously shared the stories, the cultural knowledge and information, the wisdom and perspective that has been handed down from the øl̓čmúsšn, the ancestors, from the beginning of human time. This chapter is based on that foundation.

Note

1. Information in the chapter relating to the Salish Language and to the culture and history of the Sélis (Salish or "Flathead") and Ql̓ispé (Upper Kalispel or "Pend d'Oreille") people is courtesy of the Séliš-Ql̓ispé Culture Committee, a Department of the Confederated Salish and Kootenai Tribes.

Literature Cited

Abatzoglou, John T., & Williams, A. Park (2016). Impact of anthropogenic climate change on wildfire across Western US forests. *Proceedings of the National Academy of Sciences,* 113(42), 11770–11775. doi: 10.1073/pnas.1607171113. Checked on 11 Aug 2021.

Arno, S. F. (1986). Whitebark pine cone crops—A diminishing source of wildlife food? *Western Journal of Applied Forestry*, 1(3), 92–94.

Arno, S. F., & Habeck, J. R. (1972). Ecology of Alpine larch [Larix Lyallii Parl]. In the Pacific Northwest. *Ecological Monographs*, 42: 417–450.

Arno, S. F., & Hoff, R. J. (1990). Pinus Albicaulis Engelm. Whitebark pine. In *Silvics of North America.* Washington DC: Department of Agriculture, Forest Service, 1, Conifers: 268–79. Agriculture Handbook 654 654.

Arno, S. F., & Weaver, T. (1990). Whitebark pine community types and their patterns on the landscape. In *Proceedings of the Symposium on Whitebark Pine Ecosystems: Ecology and Management of a High-Mountain Resource.* General Technical Report INT-270 (pp. 97–105). Ogden, UT: USDA Forest Service, Intermountain Research Station.

Aubry, C., Goheen, D., Shoal, R., Ohlson, T., Lorenz, T., Bower, A., … Sniezko, R. A. (2008). *Whitebark pine restoration strategy for the pacific northwest 2009–2013.* Region 6 Report. Portland, OR: USDA Forest Service, Pacific Northwest Region.

Barbero, R., Abatzoglou, J. T., Larkin, N. K., Kolden, C. A., & Stocks, B. (2015). Climate change presents increased potential for very large fires in the contiguous United States. *International Journal of Wildland Fire,* 24(7), 892. doi: 10.1071/WF15083. Checked on 11 Aug 2021.

Becker, D. M., & Lichtenberg, J. S. (2007). An update on the trumpeter swan reintroduction on the flathead reservation. *Tribal wildlife Management Program.* Pablo, MT: Confederated Salish & Kootenai Tribes. Retrieved from https://www.fws.gov/uploadedFiles/TrumpeterSwanPaper.pdf. Checked On 11 Aug 2021.

Bentz, Barbara, Logan, Jesse, MacMahaon, Jim, Allen, Craig, Ayres, Matt, Berg, Allan Carroll et al. (2009). *Bark beetle outbreaks in Western North America: Causes and consequences*. Salt Lake City, UT: University of Utah Press.

Bigart, Robert, & Woodcock, Clarence (Eds.) (1996). *In the Name of the Salish & Kootenai Nation: The 1855 Hell Gate Treaty and the Origin of the Flathead Indian Reservation.* Original printing 1996, 2nd printing 2008. Pablo, MT: Salish Kootenai College Press.

CMP (2021). Crown managers partnership. Retrieved from https://www.crownmanagers.org/. Checked on 11 Aug 2021.

CMP-Hi5 (2019). Crown of the continent high five working group technical team. Crown of the continent ecosystem whitebark pine restoration strategy: Pilot project summary. *Crown Managers Partnership.* Retrieved from https://www.crownmanagers.org/five-needle-pine-working-group. Checked on 11 Aug 2021.

Coates, Peter S., Ricca, Mark A., Prochazka, Brian G., Brooks, Matthew L., Doherty, Kevin E., Kroger, Travis, … Casazza, Michael L. (2016). Wildfire, climate, and invasive grass interactions negatively impact an Indicator species by reshaping sagebrush ecosystems. *Proceedings of the National Academy of Sciences,* 113(45), 12745–12750. doi: 10.1073/ pnas.1606898113. Checked 11 Aug 2021.

Colbourne, Rick, Moroz, Peter, Hall, Craig, Lendsay, Kelly, & Anderson Robert, B. (2020). Indigenous works and two eyed seeing: Mapping the case for indigenous-led research. *Qualitative Research in Organizations and Management: An International Journal*, 15(1), 68–86. doi: 10.1108/QROM-04-2019-1754. Checked on 11 Aug 2021.

Cooper, S. V., Neiman, K. E., & Roberts, D. W. (1991). *Forest Habitat types of northern Idaho: A second approximation.* General Technical Report INT-236. Ogden, UT: USDA Forest Service, Intermountain Research Station.

Crone, Elizabeth E., McIntire, Eliot J. B., & Brodie, Jedediah (2011). What defines mast seeding? Spatio-temporal patterns of cone production by whitebark pine: Spatio-temporal patterns of whitebark pine mast. *Journal of Ecology*, 99, 438–444. doi: 10.1111/j.1365- 2745.2010.01790.x. Checked 11 Aug 2021.

CSKT (2021). Confederated Salish & Kootenai tribes of the flathead reservation. Retrieved from http://www.csktribes.org. Checked 11 Aug 2021.

CSKT climate change strategic plan (2013). Pablo, MT: Confederated Salish and Kootenai Tribes of the Flathead Reservation. Retrieved from http://www.csktribes.org/ CSKTClimatePlan.pdf. Checked on 11 Aug 2021.

CSKT Cultural Preservation Office (2000). Cultural resource overview for the U.S. Fish and Wildlife Service western Montana management properties. Retrieved from https://csktribes.org/history-and-culture/cultural-preservation. Checked 11 Aug 2021.

CSKT Fish-Wildlife (2021). Explore the river: Bull trout, tribal people, and the Jocko River. *CSKT Fish, Wildlife, Recreation & Conservation.* Retrieved from http://fwrconline. csktnrd.org/Explore/ExploreThe-River/. Checked 11 Aug 2021.

CSKT-Annual Reports (2011). Confederated Salish & Kootenai tribes' annual report 2011. *Confederated Salish and Kootenai Tribes.* Retrieved from https://csktribes.org/government/annual-reports. Checked 11 Aug 2021.

——— (2021). CSKT annual report 2019–2020. Retrieved from https://csktribes.org/component/rsfiles/files?folder=CSKT%2BAnnual%2BReports. Checked 11 Aug 2021.

CSKT-Climate (2016). T*he confederated Salish and Kootenai tribes of the flathead reservation: Climate change strategic plan (Sep 2013, updated Apr 2016).* Pablo, MT: Confederated Salish and Kootenai Tribes. Retrieved from http://csktclimate.org. Checked 11 Aug 2021.

CSKT-CRP (1996a). Flathead reservation comprehensive resources plan volume 1 existing conditions. Retrieved from http://csktclimate.org/downloads/Comp_Plan/Comp
%20Plan%20Volume%201.pdf. Checked 11 Aug 2021.

——— (1996b). Flathead reservation comprehensive resources plan volume II policies. Retrieved from http://csktclimate.org/downloads/Comp_Plan/Comp%20Plan
%20Volume%202.pdf. Checked 11 Aug 2021.

CSKT-Forestry (2000). Flathead Indian reservation Forest management plan: An ecosystem approach to tribal Forest management. Retrieved from http://www.csktribes.org/component/rsfiles/download?path=-Forestry%252Ffmp05.pdf. Checked 11 Aug 2021.

——— (2020). Tribal forestry. Retrieved from http://www.csktribes.org/natural-resources/ tribal-forestry. Checked 11 Aug 2021.

CSKT-NRD (2021a). Explore the river – Bull trout, tribal people, and the jocko river. *CSKT's Online Educational Resources.* Retrieved from http://fwrconline.csktnrd.org/Explore/ index.html. Checked 11 Aug 2021.

——— (2021b). Fire on the land: Native people and fire in the northern Rockies. *CSKT's Online Educational Resources.* Retrieved from http://fwrconline.csktnrd.org/Fire/ index.html. Checked 11 Aug 2021.

——— (2021c). Living landscapes – Culture, climate science & education in tribal and native communities. *CSKT's Online Educational Resourc-*

es. Retrieved from https://www. skclivinglandscapes.org. Checked 11 Aug 2021.

Devlin, Sarah (2021). The Salish tribe – The Treaty of Hellgate. *Salish Tribe History*. Retrieved from https://salishtribe.wordpress.com/location-history/treaty-of-hellgate/. Checked 11 Aug 2021.

Ellison, A. M., Bank, M. S., Clinton, B. D., Colburn, E. A., Elliott, K., Ford, C. R., … Webster, J. R. (2005). Loss of foundation species: Consequences for the structure and dynamics of forested ecosystems. *Frontiers in Ecology and the Environment*, 3(9), 479–486.

Farnes, P. E. (1990). SNOWTEL and snow course data: Describing the hydrology of whitebark pine ecosystems. In *Proceedings of the Symposium on Whitebark Pine Ecosystems: Ecology and Management of a High-Mountain Resource.* General Technical Report INT-270 (pp. 302–305). Ogden, UT: USDA Forest Service, Intermountain Research Station.

Finch, Deborah M. (2012). *Climate change in grasslands, shrublands, and deserts of the Interior American West: A review and needs assessment.* Gen. Tech. Rep. RMRS- GTR-285. Fort Collins, CO: United States Department of Agriculture, Forest Service, Rocky Mountain Research Station. Retrieved from https://www.fs.usda.gov/rmrs/publications/climate-change-grasslands-shrublands-and-deserts-interior-american-west- review-and. Checked 11 Aug 2021.

Forcella, F. (1978). Flora and chorology of the Pinus Albicaulis-Vaccinium scoparium association. *Madrono*, 25, 139–150.

Forcella, F., & Weaver, T. (1977). Biomass and productivity of the subalpine Pinus Albicaulis: Vaccinium scoparium association in Montana, USA. *Vegetation*, 95–105.

Fryer, J. L. (2002). Pinus albicaulis. *Fire Effects Information System (FEIS).* Retrieved from https://www.fs.fed.us/database/feis/plants/tree/pinalb/all.html. Checked 11 Aug 2021.

Greenlaw, Suzanne, Knowlden, Smantha, Landis, Catherine, & Talucci, Tyler (2009). Biocultural restoration in an urban watershed. *Onandaga Creek.* Syracuse, NY: Onandaga Environmental Institute. Retrieved from https://oei2.org/wp-content/uploads/2018/12/biocultural_restoration_project.pdf. Checked 11 Aug 2021.

Keane, R. E., & Arno, S. F. (1993). Rapid decline of whitebark pine in western Montana: Evidence from 20-year remeasurements. *Western Journal of Applied Forestry*, 8(2), 44–47.

Keane, R. E., Holsinger, L. M., Mahalovich, M. F., & Tomback, D. F. (2017). *Restoring whitebark pine ecosystems in the face of climate change.* Gen. Tech. Rep. RMRS 361 Fort Collins, CO: United States Department of Agriculture, Forest Service, Rocky Mountain Research Station. Retrieved from https://www.fs.fed.us/rm/pubs_series/rmrs/gtr/rmrs_gtr361. pdf. Checked 11 Aug 2021.

Keane, R. E., & Parsons, R. (2010). Restoring whitebark pine forests of the northern Rocky Mountains. *Ecological Restoration*, 28(1), 56–70.

Keane, R. E., Tomback, D. F., Aubry, C. A., Bower, A. D., Campbell, E. M., Cripps, C. L. et al. (2012). *A range-wide restoration strategy for whitebark pine (Pinus Albicaulis).* Gen. Tech. Rep. RMRS-GTR-279. Fort Collins, CO: United States Department of Agriculture, Forest Service, Rocky Mountain Research Station. Retrieved from http://www.fs.fed.us/rm/publications. Checked 11 Aug 2021.

Kendall, K. C., & Keane, R. E. (2001). Whitebark pine decline: Infection, mortality, and population trends. In Tomback, D. F., Arno, S. F., Keane, R. E. (eds.) *Whitebark Pine Communities: Ecology and restoration* (pp. 221–242). Washington DC: Island Press.

Kichas, Nickolas E., Hood, Sharon M., Pederson, Gregory T., Everett, Richard G., & McWethy, David B. (2020). Whitebark pine (Pinus Albicaulis) growth and defense in response to Mountain Pine beetle outbreaks. *Forest Ecology and Management,* 457 (February), 117736. doi: 10.1016/j.foreco.2019.117736. Checked 11 Aug 2021.

Kichas, Nickolas E., Trowbridge, Amy M., Raffa, Kenneth F., Malone, Shealyn C., Hood, Sharon M., Everett, Richard G., … Pederson, Gregory T. (2021). Growth and defense characteristics of whitebark pine (Pinus Albicaulis) and lodgepole pine (Pinus contorta Var Latifolia) in a high-elevation, disturbance-prone mixed-conifer Forest in northwestern Montana, USA. *Forest Ecology and Management,* 493(August), 119286. doi: 10.1016/j.foreco.2021.119286. Checked 11 Aug 2021:

Kimmerer, Robin (2011). Restoration and reciprocity: The contributions of traditional ecological knowledge. In D. Egan, E. E. Hjerpe & J. Abrams (Eds.), *Human dimensions of ecological restoration* (pp. 257–276). Washington, DC: Island Press/Center for Resource Economics. doi: 10.5822/978-1-61091-039-2_18. Checked 11 Aug 2021.

Kimmerer, Robin Wall (2013). Returning the gift. *Center for Humans & Nature*: Expanding Our Natural & Civic Imagination (blog). Retrieved from https://www.humansandnature.org/earth-ethic-robin-kimmerer. Checked 11 Aug 2021.

Kimmerer, Robin, Wall, Brittani Orona, & Reed, Charley (2021). Practicing biocultural restoration. Presented at the 2021 Biocultural Restoration Event, Virtual. Retrieved from https://www.rrnw.org/biocultural/. Checked 11 Aug 2021.

Lanner, R. M., & Gilbert, B. K. (1994). Nutritive value of White-Bark Pine seeds, and the question of their variable dormancy. In *Proceedings of the international Workshop on Subalpine Stone Pines and Their Environment: The State of Our Knowledge* (pp. 206–211). General

Technical Report INT-GTR-309. Ogden, UT: USDA Forest Service, Intermountain Research Station.

Little, E. L. Jr., & Critchfield, W. B. (1969). *Subdivisions of the Genus Pinus.* USDA Forest Service miscellaneous Publication 1144. Washington DC: United States Department of Agriculture, Forest Service.

Luckman, B. H., Jozsa, L. A., & Murphy, P. J. (1984). Living seven-hundred-year-old Picea engelmannii and Pinus Albicaulis in the Canadian Rockies. *Arctic and Alpine Research,* 16(4), 419–422. doi: 10.2307/1550903. Checked 11 Aug 2021.

MacRitchie, Sarah (2018). Bridging western and indigenous knowledges: Two-eyed seeing and the development of a country food strategy in the Northwest Territories. *Masters, Graduate School of Public and International Affairs, Ottawa, Canada:* University of Ottawa. Retrieved from https://ruor.uottawa.ca/bitstream/10393/37576/4/MACRITCHIE,Sarah.20181.pdf. Checked 11 Aug 2021.

Mahalovich, Mary F., Burr, Karen E., & Foushee, David L. (2006). Whitebark pine germination, rust resistance, and cold hardiness among seed sources in the inland northwest: Planting strategies for restoration. In *National Proceedings: Forest and Conservation Nursery Associations – 2005* (pp. 92–101). Proceedings of the RMRS-P-43. Fort Collins, CO: United States Department of Agriculture, Forest Service, Rocky Mountain Research Station.

Marshall, Albert (2017). Two-eyed seeing – Elder albert Marshall's guiding principle for inter-cultural collaboration Presented at the Climate Change. In Presented at the *Climate Change, Drawdown & the Human Prospect: A Retreat for Empowering our Climate Future for Rural Communities,* Pugwash, Nova Scotia, Canada, October 28. Retrieved from http://www.integrativescience.ca/uploads/files/Two-Eyed%20Seeing- AMarshall-Thinkers%20Lodge2017(1).pdf. Checked 11 Aug 2021.

MCA (2017). Montana climate assessment. 2017 *Montana Climate Assessment.* Retrieved from https://montanaclimate.org. Checked 11 Aug 2021.

McCabe, B. (2021). Ghost forests. PBS. Retrieved from https://www.montanapbs.org/programs/ghost-forests/. Checked 11 Aug.

McCarthy, Maureen (2020). Native waters on arid lands. *NWAL.* Retrieved from https:// nativewaters-aridlands.com. Checked 11 Aug 2021.

McCaughey, Ward W., & Schmidt, Wyman C. (2001). Taxonomy, distribution, and history. In D. F. Tomback, S. F. Arno & R. E. Keane (Eds.), *Whitebark pine communities: Ecology and restoration* (pp. 29–40). Washington DC: Island Press.

McCaughey, W. W., & Tomback, D. F. (2001). The natural regeneration process. In Tomback, D. F., Arno, S. F., Keane, R. E. (eds.) *Whitebark Pine Communities: Ecology and restoration* (pp. 105–120). Washington DC: Island Press.

MCO (2021). Montana Climate Office *University of Montana, Forest, Montana, & Conservation Experiment Station.* Retrieved from https://climate.umt.edu. Checked 11 Aug 2021.

Minore, D. (1979). *Comparative autecological characteristics of northwestern tree species: A literature review.* General Technical Report PNW-87. Portland. OR: USDA Forest Service, Pacific Northwest Forest and Range Experiment Station.

Mitchell, S., & Kwasset, W. H. (2020). Indigenous prophecy and Mother Earth. In A. E. Johnson & K. K. Wilkinson (Eds.), All we can save: Truth, courage, and solutions for the climate crisis (pp. 16-28). New York, NY: One World.

NWAL (2020). Native waters on arid lands climate projections. Retrieved from https://native-waters-aridlands.com/climate-projections/. Checked 11 Aug 2021.

Peltier, Cindy (2018). An application of two-eyed seeing: Indigenous research methods with participatory action research. *International Journal of Qualitative Methods,* 17(1): 160940691881234. doi: 10.1177/1609406918812346. Checked 11 Aug 2021.

Perkins, Dana L., & Swetnam, Thomas W. (1996). A dendroecological assessment of white-bark pine in the sawtooth–salmon river region, Idaho. *Canadian Journal of Forest Research,* 26(12), 2123–2133. doi: 10.1139/x26-241. Checked 11 Aug 2021.

Pfister, R. D., Kovalchik, B. L., Arno, S. F., & Presby, R. C. (1977). *Forest habitat types of montana.* Gen. Tech. Rep. INT-34. USDA Forest Service, Intermountain Forest and Range Experiment Station.

Polley, H., Wayne, Derek W., & Bailey, Robert S., & Stafford-Smith, M. (2017). Ecological consequences of climate change on rangelands. In D. D. Briske (Ed.), *Rangeland systems.* Springer series on environmental management (pp. 229–260). Cham: Springer International Publishing. doi: 10.1007/978-3-319-46709-2_7. Checked 11 Aug 2021.

Resler, L. M., & Tomback, D. F. (2008). Blister rust prevalence in krummholz whitebark pine: Implications for treeline dynamics, northern Rocky Mountains, MT, USA. *Arctic, Antarctic, and Alpine Research,* 40(1), 161–170.

Robbins, C. T., Schwartz, C. C., Gunther, K. A., & Servheen, C. (2006). Grizzly bear nutrition and ecology studies in Yellowstone National Park. *Yellowstone Science*, 14(3), 19–26.

Six, D. L., & Bracewell, R. (2015). Dendroctonus. In Vega, F. C., and Hofsteller, (eds.) *Bark Beetles: Biology and Ecology of Native and Invasive Species* (London: Academic Press), pp. 305–350.

Sniezko, R. A., Kegley, A., Danchok, R., & Long, S. (2018). *Blister rust resistance in whitebark pine (Pinus Albicaulus)—Early results following artificial inoculation of seedlings from Oregon, Washington, Idaho, Montana, California, and British Columbia seed sources.* RMRS, P-76. Washington DC: USDA Forest Service.

SQCC (2021) Séliš–Q̓lispé culture committee. Confederated Salish and Kootenai Tribes. Retrieved from http:/www.cskt salish,org. Checked 11 Aug 2021.

Tomback, D. F. & Achuff, P (2010). Blister rust and western Forest biodiversity: Ecology, values and outlook for White pines. *Forest Pathology*, 40 (3-4), 186-225. doi 10,1111/j.1439-0329.2010.00655.x. Checked 11 Aug 2021.

Tomback, D. F. (1989). The broken circle: Fire, birds and whitebark pine. In T. Walsh (Ed.), *Wilderness and wildfire* (pp. 14–17). Miscellaneous Publication 50. Missoula, MT: University of Montana, School of Forestry, Montana Forest and Range Experiment Station.

——— (2018). Whitebark pine status and the potential role of biotechnology in restoration. Webinar. Committee on Forest Health and Biotechnology. Retrieved from http://nas-sites.org/dels/files/2018/03/Diana-Tomback-Presentation.pdf. Checked 11 Aug 2021.

Tomback, D. F., Anderies, A. J., Carsey, K. S., Powell, M. L., & Mellmann-Brown, S. (2001). Delayed seed Germination in Whitebark pine and regeneration patterns following the Yellowstone fires. *Ecology*, 82(9), 2587–2600. doi: 10.1890/0012-9658(2001)082[258 7:DSGIWP]2.0.CO;2. Checked 11 Aug 2021.

Tomback, D. F., & Kendall, K. (2001). Biodiversity Losses: A downward Spiral. In D. F. Tomback, S. F. Arno & R. E. Keane (Eds.), *Whitebark pine communities: Ecology and restoration.* Washington DC: Island Press.

Tomback, D. F., & Linhart, Y. B. (1990). The evolution of bird-dispersed pines. *Evolutionary Ecology*, 4(3), 185–219.

United States Census Bureau (2012). *2010 census American Indian and Alaskan native summary file.* United States Census Bureau. Retrieved from https://www.census.gov/newsroom/releases/archives/news_conferences/20121210_aian_webinar.html. Checked 11 Aug 2021.

United States Department of Agriculture (1975). S*oil taxonomy, A basic system of soil classification for making and interpreting soil surveys.* Agriculture Handbook 436. Washington DC: United States Department of Agriculture.

USDA-NRCS (2019). PLANTS database, plant list of accepted nomenclature, taxonomy, and symbols. Retrieved from https://plants.usda.gov/home. Checked 11 Aug 2021.

United States Department of Agriculture-USFS (2021). White pine blister rust. *High Elevation White Pines*. Retrieved from https://www.fs.fed.us/rm/highelevationwhitepines/Threats/blister-rust-threat.htm. Checked 11 Aug 2021.

United States Geological Survey (2021). Predicting climate change impact on fish. Retrieved from https://www.usgs.gov/special-topic/drought/science/climate-change-impacts-fish ?qt-science_center_objects=0#qt-science_center_objects. Checked 11 Aug 2021.

White, Germaine, & CSKT (2010). *Fire on the land: Native peoples and fire in the northern Rockies: Salish and Kootenai tribes fire History Project.* Lincoln, NE: University of Nebraska Press.

Whitlock, C., Cross, W., Maxwell, B., Silverman, N., & Wade, A. A. (2017). *2017 Montana climate assessment*. Bozeman & Missoula, MT: Montana State University and University of Montana, Montana Institute on Ecosystems. Retrieved from http://montanaclimate. org/chapter/title-page. Checked 11 Aug 2021.

Zhao, Chuang, Liu, Bing, Piao, Shilong, Wang, Xuhui, Lobell, David B., Huang, Yao et al. (2017). Temperature increase reduces global yields of major crops in four independent estimates. *Proceedings of the National Academy of Sciences,* 114(35), 9326–9331. doi: 10.1073/pnas.1701762114. Checked 11 Aug 2021.

Index